WARM-UP EXERCISE

Through the medium of the printed page, you are going to witness engineers in action. The first unit of this learning program will give you a general impression of what engineers do, the problems, challenges, and difficulties they encounter, the qualities they must possess to perform their function, and the rewards that are available to them. As you proceed through the learning program, you will be asked to respond to exercises designed to give the author a chance to interact with you much in the same way he would if he could meet with you personally to talk about engineering. Please get into the spirit of this interaction right from the beginning by responding to this Warm Up exercise with pen or pencil in hand.

Assume that you are the president of an engineering consulting firm. A group of state and local officials have requested that you solve the following problem for them.

> We have this very large and long bay. The mouth of the bay is 29 kilometers at its narrowest point. We need a way for vehicular traffic to motor across the mouth of the bay rather than taking the detour around the entire bay.
>
> It seems feasible to build a trestle type of bridge across the entire stretch of open water, but we cannot have the shipping lane vulnerable to being blocked by any collapse of the bridge into the water, such as might be the case in the event of an enemy attack or sabotage.
>
> Yet we cannot afford to tunnel the entire 29 kilometers, and ferrying traffic is not satisfactory either.
>
> Can you solve this problem for us?

How would you approach a solution to this problem? A picture essay on the solution appears in the following pages.

THE CHESAPEAKE BAY BRIDGE-TUNNEL

The numerous trestles, bridges, and tunnels of the $140 million structure in Figure 1 (a and b) carry vehicles across the 29-kilometer mouth of Chesapeake Bay. The bridge-tunnel is a remarkable engineering achievement. It is the longest man-made fixed crossing of navigable ocean, built to withstand punishing waves and rushing tides.

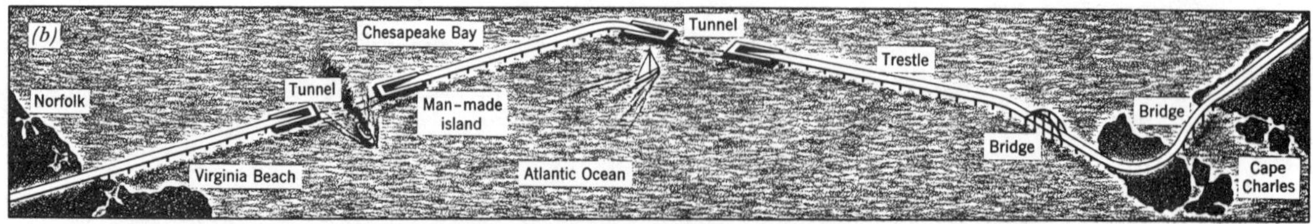

Figure 1. (a) View of the Chesapeake Bay Bridge-Tunnel looks south from Cape Charles toward Virginia Beach, with the Atlantic Ocean at the left, the bay at the right. (b) Starting from Virginia Beach, a vehicle moves over 5 kilometers of concrete trestle and then, at the first of four man-made islands, enters a 1.6-kilometer tunnel that carries traffic beneath a major shipping lane. The vehicle continues over 6 more kilometers of trestle, through another tunnel, then over more trestle, two bridges, and a natural island before it reaches the mainland at Cape Charles.

As is often the case, this span was designed by an engineering consulting firm that specializes in such projects. The firm was commissioned to locate, design, and supervise construction of the entire structure. The unusual restriction imposed on this structure (that it could not pass over the main shipping channels, because a bridge above them could be destroyed and possibly trap navy vessels in the bay) made it necessary to go under the channels, employing tunnels where bridges would have been placed under ordinary circumstances.

Figure 2. Major components of the 19-kilometer trestle: concrete piles, concrete cross beams, and concrete slabs that form the roadway.

Figure 3. In the foreground is a man-made island that provides protection for the tunnel where it surfaces to connect with the trestle. The island at the other end of this tunnel is visible in the background. Each island is about 500 meters long and required almost 2 million tons of material. The island's sand core is protected by a heavy armor of stone.

There are, of course, many extremely complex problems to be solved in designing and executing a project such as the Chesapeake Bay Bridge–Tunnel. Can you list just a few of these? (How, for example, will the trestle piles, cross members, and roadway slabs be constructed most economically?)

____

____

Will any new types of equipment be needed?

____

Engineers must give careful attention to the means by which their creations will be built. In a case like this, the construction methods are as important as the design itself. In particular, you will notice that trestle piles, cross members, roadway slabs, tunnel sections, and other components were prefabricated by mass production methods on land, where construction is achieved with less difficulty and expense. The use of prefabricated components is a method of construction resulting from adequate forethought on the part of the structure's designers.

The interrelationship between a structure's physical characteristics and the means of constructing it is also apparent if you look at the special construction equipment used. Determination of what equipment was needed and the design of it were major parts of the problem. In fact, the investment in construction equipment for this project was $15 million. Thus, the feasibility of such a venture depends heavily on the engineers ability to formulate a design that minimizes construction costs while satisfying all functional requirements.

Figure 4. The trestle was constructed by driving long (up to 52 meters) concrete piles deep into the bed of the bay by means of a specially designed machine nicknamed Big D, shown in (a) and (b). This $1.5 million rig is an enormous pile driver and crane mounted on a barge with retractable legs.

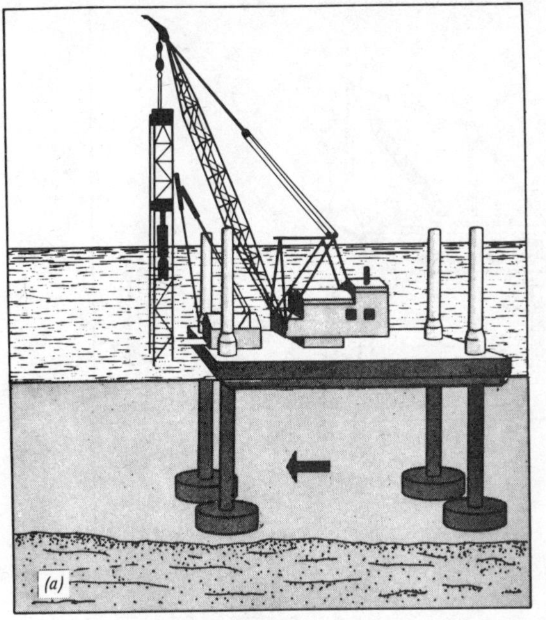

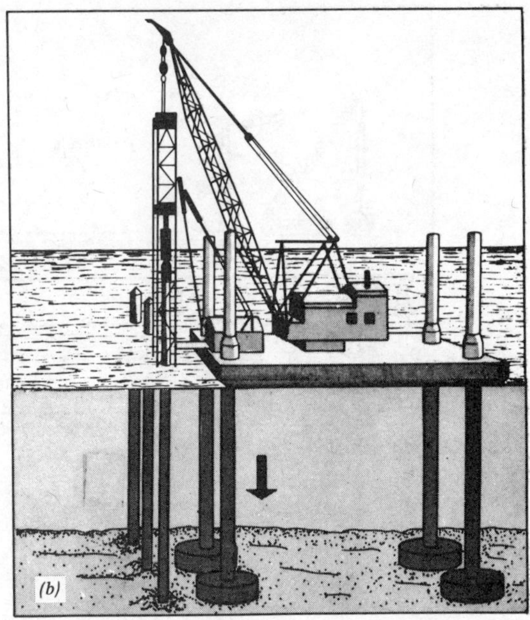

Figure 5. Big D floats (a) into position and then becomes a stationary platform by extending its legs and hydraulically lifting itself out of the water (b). It "stands" in this position until its work is complete at that spot. Without this stability Big D would never be able to locate the huge piles to close tolerances.

Figure 6. Big D is followed by another special rig called the Two-Headed Monster, shown in (a). As it crawls along the tops of the piles, the leading "head" of the Monster trims the piles to a uniform height. Cross beams are placed by the Monster's trailing "head."

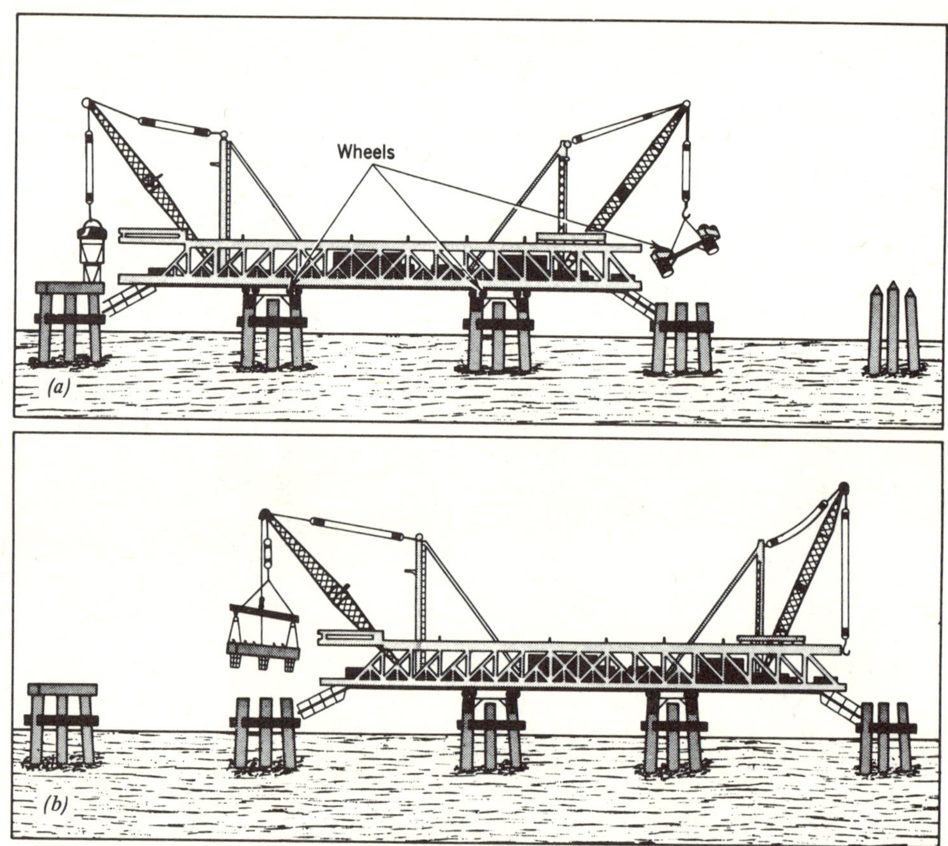

Figure 7. The Monster moves along on wheels that are temporarily mounted on the tops of the piles. It has an extra set of wheels that it places on the next row of piles ahead of it (a), after which it rolls forward to its next working position (b).

Figure 8. Then this machine, called the Slabsetter, places four slabs of precast concrete side-by-side to form the roadway. Slabsetter moves along by using two sets of tracks (girders).

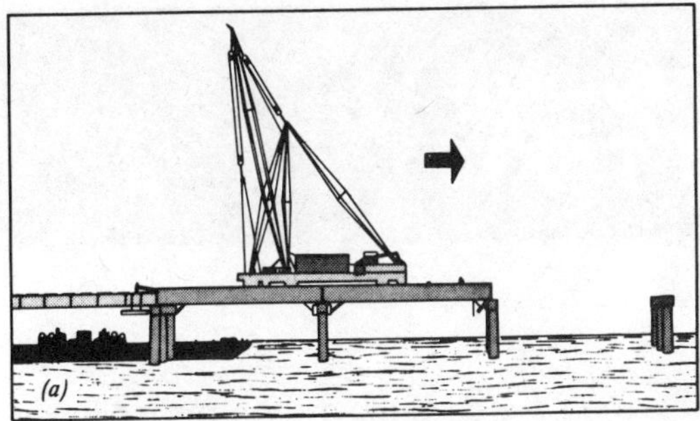

Figure 9. After finishing its work on the span at the left (a), Slabsetter rolls onto the alternate set of tracks. (b) It swings the other set of tracks around behind it and is ready to start work on the span just vacated. In this manner, it travels "hand over hand" along the tops of the piles and is on solid footing while performing its work, without ever "getting its feet wet."

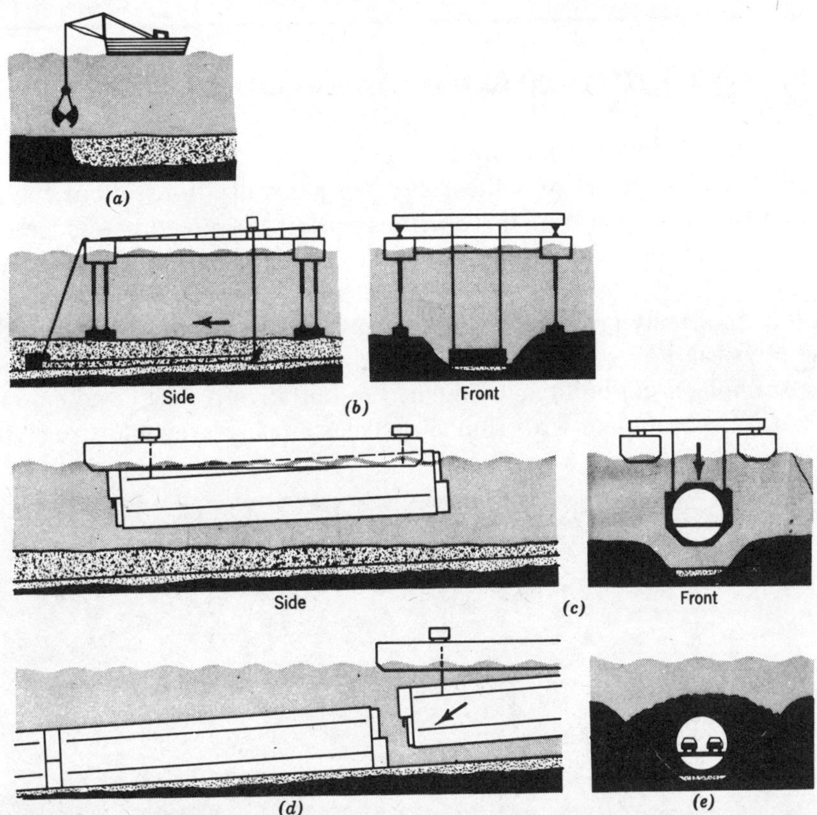

Figure 10. Construction of the tunnels beneath the shipping channels is another interesting story. (a) First a dredge cut a trench roughly 30 meters wide and anywhere from 15 to 30 meters deep. (b) Then a device called a screed graded the bottom of this trench to within a few centimeters of the specified level. Note that the screed can be pitched to parallel the desired slope of the finished trench. (c) Next, a 90-meter prefabricated tube was lowered and positioned by a special barge, shown here. (d) It was then connected to the previously laid tube section. (e) Later the tubes were covered with layers of rock and sand, the joints sealed, the temporary end-seals removed, and the interiors finished. The result was one continuous submarine tube linking the man-made anchor islands.

Learning Segment 1
WHAT IT TAKES

LEARNING OBJECTIVES

- Describe some of the skills and attitudes that must be developed by a professional engineer.
- Increase an awareness of your strengths and weaknesses in these skills and attitudes.
- Renew your efforts to develop these skills and attitudes.

CREATING SPECIFIC SOLUTIONS TO BROADLY DEFINED PROBLEMS

What is required to design devices, machines, and structures such as those involved in the Chesapeake Bay Bridge-Tunnel? From the following generalizations about the problems engineers solve, you can infer many of the qualities possessed by competent practitioners. In this respect, all the case studies included in this learning segment will also be useful.

An engineering problem usually begins with the recognition of a need that apparently can be satisfied by some physical device, structure, or process. At this stage, the problem is likely to be vague. For example, a shipbuilding company has tentatively decided to market an air-cushion vehicle (ACV) to compete with similar vehicles already available from other companies (Figure 11).

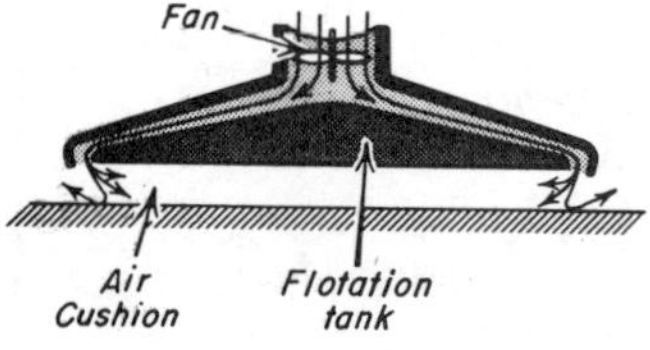

Figure 11. An ACV is supported by a cushion of air trapped within its flexible skirts, enabling it to travel over land and water with ease. The vehicle shown carries 265 passengers and 30 automobiles, cruises at 100 kilometers per hour, and can negotiate 3-meter waves. Craft of this type provide ferry service across the English Channel.

In broad terms, the management has decided on desired features of this vehicle, such as the price range, seating capacity, and cruising speed. The engineers now have their assignment: design a vehicle that satisfies the general performance characteristics given. This is typical of engineering assignments. The engineer is given the function or purpose to be served and probably some requirements and preferences for a solution. Such functional and performance specifications are usually selected by the engineer's superiors or clients, often in collaboration with the engineer. The challenge to the engineer is to translate the loose statement of what is wanted into the specifications for a satisfactory means of fulfilling that objective.

In most cases there are numerous ways of achieving the specified objective; many, if not most of them, are not obvious, perhaps not even known, at the start of the project. It is up to the engineer to uncover and explore a number of the possibilities. The knowledge gained through education and experience is an important source of such solutions, but not the only one; the engineer must also rely heavily on ingenuity.

CASE STUDY: AN AUTOMATIC PRODUCTION MACHINE

A large telephone system includes millions of switches. Engineering has made it possible for those switches to function for years, making billions of connections without wearing out. But they are still subject to failure caused by moisture or foreign matter. The cost of isolating and remedying these failures plus the cost of preventive maintenance have concerned the company for years. An engineer was asked to reduce these costs by improving the reliability of the system.

During his investigations, he developed and evaluated many possibilities; finally, he selected the ingenious switch described by Figure 12 as the most promising solution. This device is remarkably fast, very reliable, maintenance-free, and superior to any previous switch.

A very important question remains, however. The answer will determine whether this novel switch can be useful to the company and its customers. Can it be economically manufactured by the millions? To answer this question, a team of engineers was assigned the task of developing, if feasible, an economical method of making these switches.

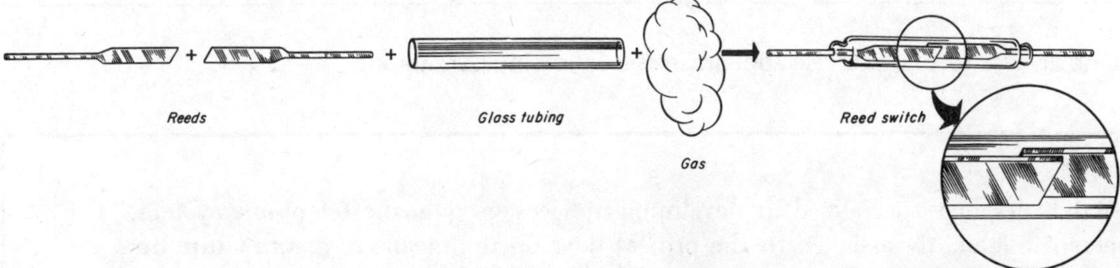

Figure 12. The guts of this switch are two metal reeds sealed in glass tubing along with an inert gas. A close-up view shows the contact point where reed alignment and spacing are critical. In use, this assembly is mounted in an electromagnetic coil that, when energized, causes the reeds to make contact and complete a circuit.

THINK IT THROUGH

Things are a little vague at the beginning of this engineering project. The solution is not even in sight. However, one expectation for this solution, and for every engineering problem's solution, is clear to the engineers. What is this very clear and universal expectation?

One answer to this exercise appears in the following paragraph.

Throughout this project, its economic feasibility was under scrutiny. The team periodically paused to reevaluate the probability of developing an economical solution. If, at any time during the project, it appeared that all methods of manufacture would be too costly, the originator of this switch would have been looking for anther solution: a switch cheaper to manufacture.

THINK IT THROUGH

What do you suppose is the main difference between the switch designer and the engineers subsequently assigned to the project?

One answer to this exercise appears in the following paragraph.

The switch designer specialized in developing devices used in the telephone system. The engineers subsequently assigned to the project develop the means to manufacture these devices. They are known as *process* or *manufacturing engineers*. Members of such design teams are usually experts in complementary fields. In this particular case, one engineer specializes in the behavior and forming of glass, another in machine components and mechanisms, another in electrical and magnetic phenomena, and so on. A close working relationship among members of an engineering team is vital, since each specialist concentrates on a facet that interacts with most other facets of the solution. One member of the group (referred to as the *project engineer*) serves as a coordinator of the group's activities, so that all parts of the final system are properly interrelated.

TECHNICAL TERMS

Precise meanings should be assigned to technical terms. This exercise is designed to give you the opportunity to fix in your mind the meanings of some of the terms that have been introduced. Write a brief definition for each term.

1. Process engineer (include one synonymous term)

____

____

____

2. Project engineer

____

____

____

1. A process engineer, also called a manufacturing engineer, specializes in developing the means to manufacture devices in relatively large numbers.

2. A project engineer, a member of the engineering team, coordinates the activities of the team.

THINK IT THROUGH

Briefly outline this engineering team's problem in relation to the telephone switch.

____

____

____

____

An answer to this exercise appears in the following paragraph.

This team's problem was to find the most economical means of transforming glass tubing, reeds, and gas into the specified switches—by the millions. It didn't take the team long to realize that making millions of switches by hand would be prohibitively expensive. The fate of this clever switch hinged on the team's ability to develop an economical machine—or a combination of operator and machine—to manufacture it. Of course, developing a method that places these reeds in glass tubing and aligns them to the close tolerances required is a challenge.

THINK IT THROUGH

Does the page from the engineer's notebook (Figure 13) tell you anything about the way solutions to engineering problems are worked out? Does it give you any clues to the way this switch manufacturing problem will finally be solved?

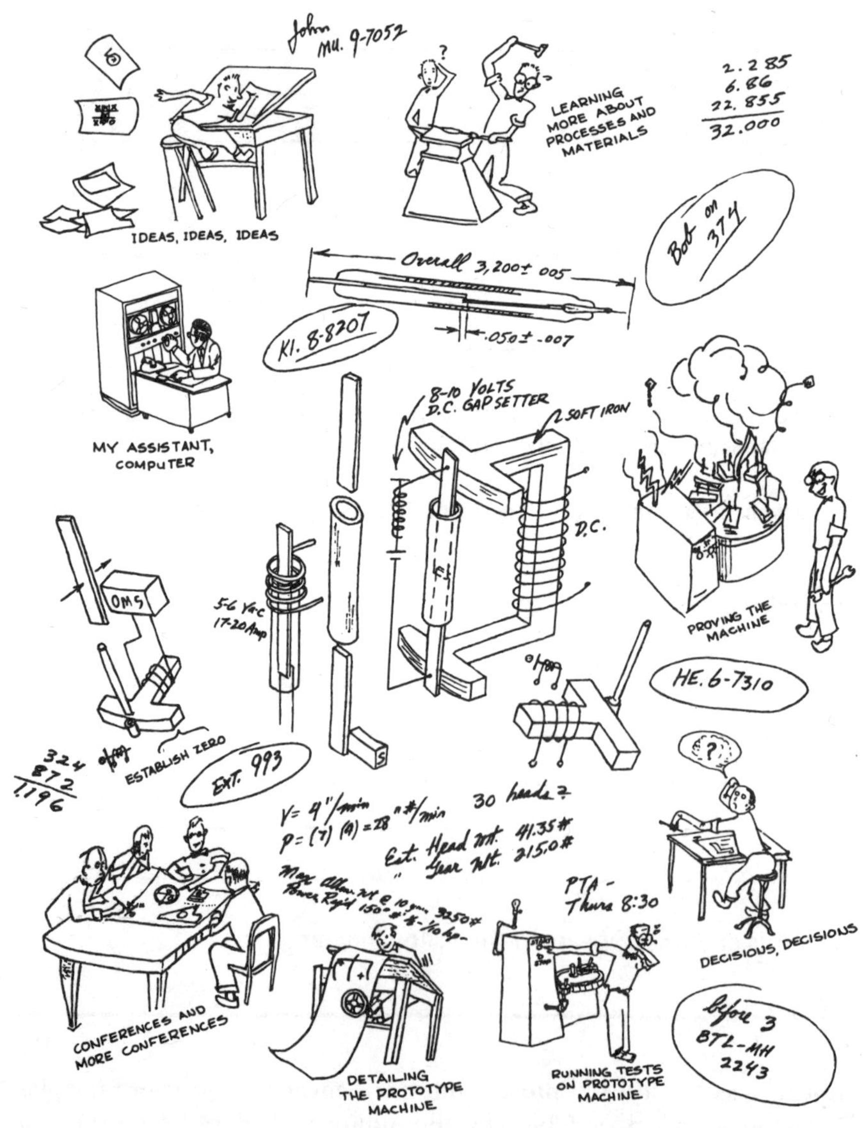

Figure 13. A page from an engineer's notebook

Some possible answers to this exercise appear in the following paragraphs.

The page from the notebook of one of the team's engineers reflects, by way of his doodlings, the many alternatives, false starts, details, procedural steps, and communications involved in the development of the machine shown in Figure 14.

When the team believed that it had devised the most economical method, its proposal had to be specified in detail so that technicians and craftsmen could construct a prototype. The engineers were responsible for overseeing the construction of this prototype. They made modifications of their original design during the construction. When the model was complete, they supervised test runs of the machine; as a result, additional design modifications were instituted. Finally, after much testing and refining, the proposed machine was ready.

Figure 14. An automatic machine that makes more than a million reed switches a year. It operates on the merry-go-round principle. The turret containing 18 identical assembly head revolves and, as it does so, the reed switch gradually takes shape. At sequential stations around the turret's periphery, tubing is placed, reeds are inserted and aligned, the gap is set, gas is injected, the tubing is sealed, and the finished switch is ejected. The switches then go to the testing section of the machine where their physical and electrical characteristics are measured. Unsatisfactory switches are rejected and, on the basis of these measurements, the machine adjusts itself to correct whatever is causing the defects.

Complete specifications of the final prototype were prepared by draftsmen so that additional machines could be constructed. Then a more effective switch became available for general use at a rate of many millions per year.

The task still was not complete. The process engineers followed up by observing the machine in use, recommending appropriate design changes, and evaluating the design so that future projects could benefit from experience with this machine.

FIGURE IT OUT

Why do you suppose all this engineering effort was necessary for such a simple device as this telephone switch?

____

____

____

____

____

One answer to this exercise appears in the following paragraph.

Why was this intensive engineering effort necessary for such a simple device? The simplicity is deceiving; it leads you to greatly underestimate the effort, ingenuity, analytical work, and investigation that went into this device. Without this work, the result would probably be more complicated and therefore more impressive to the untrained eye—but no more effective. In fact, a more complicated machine would be more susceptible to breakdown, more costly to make, and perhaps too expensive to sell.

ACTING UNDER URGENT DEADLINES

An air of urgency surrounds almost all engineering projects. Even if no target date has been set for a solution, there usually is pressure to produce results as quickly as possible. Consequently, the engineer generally must recommend a solution long before he has had time to uncover all the possibilities. Furthermore, instead of an exhaustive evaluation of the alternatives (something he simply does not have time for) he must rely heavily on judgment. This is a demanding aspect of an engineer's day-to-day work.

FIGURE IT OUT

In almost every instance, the engineer must propose a solution to a problem in a rather limited period of time. What do you imagine are the consequences of this restriction? (For example, what must he resort to, what must he forego?)

In general, pressure to get the job done means settling for results and resorting to methods that fall considerably short of perfection. There is not the time to conceive of all possible solutions. Exhaustive experiments to evaluate each alternative solution are out of the question; judgment must be employed where, with more time, more thorough experiments would be performed. Estimates and approximations often serve in lieu of time consuming, exact measurements and calculations.

WORKING OUT CONFLICTING OBJECTIVES

In most engineering problems there are conflicting objectives. The designer of the tracking antenna, pictured by Figure 15, knows that very well.

Figure 15. In the design of this antenna, which is used to track and communicate with space vehicles, a number of conflicting objectives had to be resolved.

Among the conflicting objectives the designer had to deal with were two major ones: long range and precise tracking ability. Very early in the project, it became evident that an attempt to increase the antenna's range by enlarging the reflector would decrease its ability to stay on target because wind-caused vibration would be increased. True, the vibration could be overcome by stiffening the reflector with more steel, but that would add to its inertia which in turn would aggravate the aiming problem. Therefore, the designer had to make some difficult compromises in order to achieve a satisfactory balance between range and aimability.

And so it goes with every problem; as the engineer attempts to achieve "the ultimate" solution with respect to one performance characteristic, he is likely to lose ground with respect to other objectives. In the end, a balance must be achieved, which is rarely a simple task.

CHECK IT OUT

Take a familiar device or structure and attempt to identify some of the conflicting objectives with which the designer presumably had to cope as he arrived at that solution. (For example, in most cases, the designer has to resolve this conflict: maximize the number and effectiveness of functions the device performs but minimize the cost of manufacturing it.) For this purpose you might select a camera, a household appliance, or a building.

____

____

____

____

____

____

The responses to Check It Out exercises depend on your choice of areas of your experience and you must therefore be the judge of your answers.

WORKING WITHIN ECONOMIC RESTRAINTS

The extent to which economic factors enter into engineering work cannot be overstated. If an engineer's creations are to benefit society, the intended users must be able to afford them. Therefore the engineer must have a keen interest in the costs of developing, producing, distributing, and using his solutions. Consider production cost—it depends mainly on the designer; he had better be concerned about it. The desalinator's ultimate cost to the consumer and, in fact, the economic success of the whole venture depend heavily on the producibility of the engineer's design. So it is with all engineering creations.

The engineer's concern with production and other costs is not solely for the benefit of the ultimate buyer. The survival of many businesses depends on the cost consciousness of the engineers. Moreover, whether or not an engineer's project ever gets off the ground depends on the economic promise of the basic concept. Private enterprise does not ordinarily embark on a venture unless it promises an attractive profit. In public ventures, too, a satisfactory benefit-to-cost ratio is demanded. Even though an engineer's proposal may satisfy its intended function completely, it is doomed if it will not yield a net gain to the business or community.

CASE STUDY: AIRCRAFT DEVELOPMENT

For four years a design team has been developing a new airplane called a VTOL (pronounced *veetall*, meaning vertical takeoff and landing plane). The results of their efforts are described in Figures 16 through 20.

Figure 16. A VTOL hovering in a stationary position. This plane can move straight up and down or hover like a helicopter by using the three lift fans—one in each wing and one in the nose. Once airborne, the pitch of louvers under the wing fans is changed to deflect the fan exhaust rearward, thus producing horizontal acceleration. When the aircraft speed is sufficient for normal wing-supported flight, the fan louvers are closed and the craft functions as a conventional, highspeed jet plane, cruising at about 800 km/hr. The fans are driven by the same two jet engines that propel the plane in horizontal flight.

THINK IT THROUGH

Two conflicting objectives should be apparent to you in the VTOL engineering project. What are they?

____

____

Answers to this exercise appear in the following paragraph.

The objective to take off and land vertically or to hover certainly would create a design conflict with the objective of high forward speed. The team has developed a plane with sufficient thrust to raise it vertically and to propel it horizontally at competitive speeds. This is a challenge if you don't want the plane to be all engine. Another challenge arises from the tendency of such planes to tilt when in a hovering position, and to drift horizontally. To overcome these tendencies, a sophisticated control system is needed to maintain the plane's hovering stability.

Figure 17. A multiple-exposure photograph of the VTOL taking off, showing the transition from vertical to horizontal flight.

The team experimented extensively with various models (Figures 18 and 19) in order to evaluate alternative designs. These engineers are skilled in instrumenting a model to obtain the necessary measurements. They also are capable of planning efficient experiments and intelligently interpreting the data that the experiments yield.

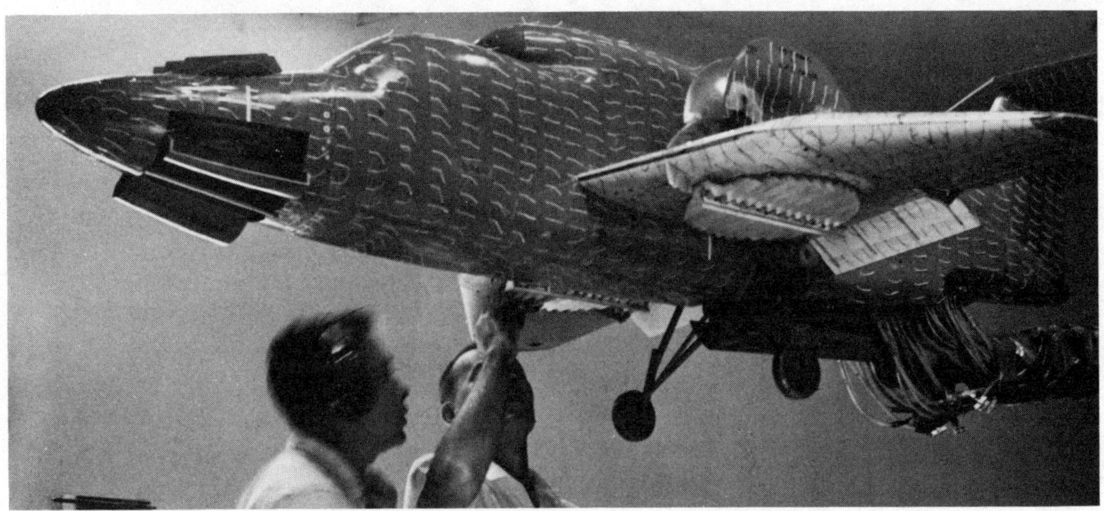

Figure 18. A one-sixth scale model of the VTOL being readied for wind tunnel tests. The tufts of wool attached to the fuselage show airflow. The cables leading into the rear of the model bring power in and information out from measuring instruments.

Figure 19. This is an experimental model of the type used in outdoor tests of preliminary VTOL designs. In this case the designers are evaluating the hover and control capabilities of a plane with three small jet engines at each wing tip. Eventually they removed the "leashes" and allowed this rig to rise to a prudent altitude. Crude as this flying framework may seem, it told the engineers what they needed to know—without unnecessary fanciness, time, and expense.

Even though working models have been testflown for many hours, the company still has no definite customer. The company has no guarantee of ever selling a VTOL. What the company receives in the future for its efforts depends on how well these seven engineers do their job and on competitors' designs.

THINK IT THROUGH

We know that engineers must work within economic constraints. Yet, the companies employing the engineering teams to develop competing VTOL aircraft may never sell any. How can this be?

One answer to this exercise appears in the following paragraph.

Ventures like this are common. A company sees a future opportunity in an engineering creation that presently is beyond the state of the art. It develops its technical competence in the area by assigning engineers to design one or more experimental models. Such development efforts are sometimes funded by the government; at other times, the cost is borne by the company in the hope that the investment will lead to profitable contracts someday. If no contracts arise, the company may be able to recover its investment through commercial sales. Figure 20 indicates that the team is already thinking of this possibility.

A company usually carries on such projects at the same time that their competitors are working on similar projects. In each instance, several companies are building technical competence, gambling that it will eventually pay off. This adds excitement to the project, but it also affects job security.

These cases provide illustrations for the generalizations, descriptions of method, discussions of impact, and other concepts explored in this program. But we are not portraying the practice of engineering through the case studies shown here. Since engineering is multifaceted, it is inevitable that there are roles and activities not represented here.

Figure 20. The stakes are more than military contracts; naturally the company hopes to penetrate the commercial market with a VTOL. Therefore the design team has been considering transport VTOLs like this one, based on the concepts that they have developed for the military version shown in Figure 16.

THINK IT THROUGH

Many engineering projects require the combined efforts of two or more engineers. A project such as VTOL development will require dozens. Identify some of the benefits as well as adverse consequences of such group efforts for the engineer (in the short and long run), for the enterprise, and for the solutions such teams create.

WORKING WITH PEOPLE AND THEIR PROBLEMS

Most of the creations described in this program are complex systems; they involve many components, and the interrelationships between them are complicated. Although it is not particularly large, the VTOL is a very complicated mechanism. This is a conspicuous trend in engineering creations. They are becoming more complicated, larger-scale, and more ambitious. Because of this complexity and the broad range of knowledge required in such projects, many engineering problems are handled by teams of engineers with varying backgrounds. The situation in which one engineer completely designs a device or structure is becoming rare. In fact, hundreds of engineers are involved in the design of a spacecraft. They are divided into teams with one team designing the propulsion subsystem, another the guidance subsystem, and so on for more than a dozen, major subsystems.

Generally, the result of an engineer's efforts is tangible (a physical device, structure, or process) as illustrated by the desalinator, the reed switch machine, the bridge-tunnel, and the VTOL. This fact is probably the basis of a common misconception about engineering. Since the result is a device, structure, machine, or mechanism, many people conclude that engineers spend most of their time working with these things (like the mechanic, the television repairman, or the laboratory technician), but this is not ordinarily so. For instance, as engineer usually does much of his problem solving in the abstract. He works much more with information (e.g., fact gathering, computing, thinking, and communicating) than with things. Furthermore, technicians are usually employed to construct models for testing and demonstrating the engineer's creations, so he has little occasion to "work with his hands." In this respect, engineering work is quite different from what many people imagine it to be.

This is true for another reason: the typical engineer spends more time with people and people-problems than you might think. The frequent use of teams and the amount of specialization that prevails in practice are two reasons that contact with people requires more of an engineer's time (and sitting in solitude at the drawing board less) than you probably realize. There are those days when most of his time is spent making inquiries, issuing instructions, answering questions, providing advice, exchanging ideas, and seeking approvals. Consequently, the ability to mesh with other people in a variety of capacities is important; so is the ability to communicate effectively.

The engineer's involvement with people does not end here. A significant part of his work is the detection and appraisal of human needs and desires. He must also be concerned with public acceptance of his solutions and, therefore, must get to know how people will use his creations, how they will react to them, and the features they prefer. He must also anticipate and be concerned about the effects of his creations on people. Thus the engineer is deeply involved with social needs, as well as with social acceptance and social effects of his creations.

LET'S BE SURE ABOUT THIS

Give at least six reasons why it would be incorrect to say that engineers need not work with people and their problems.

____

____

____

____

____

____

____

Engineers must work in teams on complex projects and must be able to deal effectively with other team members.

Engineers must be able to gather information from many different people.

Engineers must give instructions to many different people.

Engineers must answer questions for many different people.

Engineers must give advice to many people.

Engineers must exchange ideas with many people.

Engineers must seek approvals from many different people.

Engineers must win public acceptance of their designs.

Engineers must anticipate and check on how people use their designs.

Engineers must find out what features people prefer in their designs.

Engineers must anticipate and explore the effect of their creations on people.

AN OVERVIEW OF WHAT ENGINEERING TAKES

Skills

What skills are required to solve problems such as those we have discussed? First, an engineer needs problem-solving ability. This is only one of many skills that he must acquire. We have selected three of these skills for discussion in Units II and III: problem solving, modeling, and optimization.

These are not the only skills, however. Among other things, an engineer should know how to *prepare*, *conduct*, and *evaluate experiments* that will give a maximum amount of information with a minimum of time and expense. In experimentation and other phases of this work, he will rely heavily on *measurement* skills. Closely related to measurement and experimentation is the ability to *reach intelligent conclusions from observations*; skillful interpretation of them is not as straightforward as one might think. This is true because of the uncontrollable variations in the characteristics of all materials, objects, and devices; also, *no* measurement system is perfect; furthermore, most of the engineer's conclusions must be based on relatively small numbers of observations. These circumstances complicate the process of drawing conclusions. Therefore, an engineer must *learn the potential sources of error* in reaching conclusions, the limitations of small samples, the roles of chance, uncertainty, and prejudice, and the importance of carefully evaluating the reliability of available evidence. Engineers have an expression for it: "the art of skillful approximation." Skill in the *use of information resources* (books, reports, articles, patents) is becoming more and more important for engineers. With knowledge expanding so rapidly, the desirability and the difficulty of searching for know-how relating to a specific problem increase. Without this searching skill, an engineer can miss a lot of valuable information and waste time in the process.

Thought skills are obviously important in engineering, and so a major goal of engineering education, both in college and continuing professional study, is to sharpen *reasoning* and *analytical* and other *mental abilities*.

Do not underestimate the value of skill in *verbal communication*. Eventually, on-the-job experience will convince you that this is a crucial skill.

An engineer also must be skilled in mathematics. It is an important skill that is appropriately emphasized in this learning program in the segment on modeling. It will be needed most in modeling alternative solutions in order to predict their behavior.

Graphical skills provide an important means of expression, especially in engineering, where drawings, sketches, and graphs are so useful. Graphs, for example, are often used to communicate concepts, to aid thinking, and to detect as well as describe relationships between variables. The ability and inclination to think and communicate graphically are developed throughout an engineering education.

Naturally an engineer will rely extensively on *computational skills*, especially the ordinary pencil-and-paper type. But many computational tasks in the real world are so lengthy that faster computational means, such as computers, are necessary. Therefore, an engineer must develop competence with an array of computational tools suitable for jobs ranging from simple one-shot calculations to computational tasks involving millions of repetitive operations.

Another skill needed is the ability to work effectively with other people. Also needed is the ability to work within economic restraints, resolve conflicts among objectives, work within deadlines, and to turn broad problem definitions into specific solutions.

CHECK IT OUT

The following check list will help you form an inventory of your strong skills areas and of areas that you might like to improve. Write a brief evaluation of your strengths and weaknesses in each of the following skill areas.

Formulating specific solutions for broadly defined problems

Completing work within deadlines

Balancing conflicting objectives

Working within economic constraints

Working with people and their problems

Computation, including computer utilization

Graphics

Verbal communication

Mathematics

Reasoning and analytical skills

Using information resources

Finding potential sources of error

Reaching good conclusions from observations

Preparing, conducting, and evaluating experiments

Problem solving

Since this exercise is directly related to your own personal experience, there is no way to provide a correct or preferred answer. You must use this type of exercise to tell yourself something about yourself or your work. Hopefully, this learning program will help you improve in those areas you have noted as needing improvement.

Knowledge

An engineer must also possess a wealth of factual knowledge in areas such as *science*, *engineering systems*, and *economics* in order to design complicated machines and structures. In Unit IV, you will examine samples of this knowledge in detail, and this will give you further insight into engineering *and* information that you will find useful.

Attitudes

Certain qualities you bring to bear on problems involve neither factual knowledge nor skills. They constitute what can be described as attitudes. They are difficult to define, but certainly include a *questioning attitude*. This curiosity for the "how" and the "why" of things leads to much useful information and many useful ideas. Some of this questioning results from *inquisitiveness*, but some arises from a certain *skepticism* that prompts the engineer to challenge various "facts," requirements, and features to make them "prove themselves."

In such a dynamic field, *flexibility* (receptivity to changing concepts, innovations in technique, and new ideas) is obviously desirable. Furthermore, working in a dynamic field has another and related consequence: a *favorable attitude toward continual learning* is of utmost importance. This calls for a certain humility—recognition that a college diploma symbolizes the end of the beginning of a lifelong process. The willingness to continue learning is a major determiner of success in any profession.

Clearly, however, one's *attitude toward people* (employers, colleagues, and especially the people directly and indirectly affected by one's professional activities) is the most important factor. This attitude should be consistent with what is traditionally expected of professionals. The professional person serves society as an expert with respect to a type of relatively complicated problem. The layman trusts the professional person and, because of this, the professional has an obligation to perform services ethically. Since most professional activities will affect the well-being of many people, the public trusts that designs will be safe and otherwise beneficial to the public welfare. The public also trusts that it will receive the full measure of service for which it pays.

An engineer's professional obligations involve more than living up to this trust. They include:

- Insistence on seeing a project through to successful implementation of a solution.

- A desire to follow up on that solution in order to benefit from experience with it.

- A feeling of responsibility to colleagues expressed in actions, attempts to improve the status of the profession, and a willingness to exchange information.

- Maintaining in strictest confidence the unpatented ideas, secret processes, and unique know-how that one's employer or client considers proprietary.

- A concern for the direct and indirect consequences of solutions.

Granted there are people with no special training who are clever enough to design tools and other devices (but not long-span bridges or communications satellites). Granted, too, that their creations might suffice as long as they are not to be mass produced, do not see heavy use, and do not jeopardize life and limb, and as long as one insists on high reliability at low cost. Almost anyone could develop a method of fabricating reed switches, although contact alignment and spacing would be a challenge. But there is a big difference between putting a few devices together at a workbench with a bunsen burner and some hand tools and manufacturing millions of devices where thousandths of a cent and thousandths of an inch count! Only a professional can create complex physical contrivances for society *and* meet the demands for low cost, high reliability, long life, large volume, and safety. To perform their function effectively, engineers must possess certain qualities that align rather nicely into three major categories.

LET'S BE SURE ABOUT THIS

What three categories of qualities important for professional engineers have we discussed?

____

____

____

____

____

The three categories are skills (for instance, design, measurement, and communication), knowledge (of physics, history, economics, feedback systems, etc.), and attitudes (objectively, social concern, willingness to continue learning, and the like). The remainder of this learning program elaborates on certain of those skills, types of knowledge, and attitudes.

If you are discouraged by what may seem like an imposing set of qualities, do not overlook the fact that these qualities come partly from college education and partly from experience as a practicing engineer. They are qualities that must be developed and perfected over years of professional practice.

SELF QUIZ

1. An engineer receives assignments in which the problems are somewhat vaguely defined. How must the engineer's solutions differ from the vagueness of the assignment?

2. Define the term *process engineer*.

3. What is a *project engineer*?

4. What economic constraints does a process engineer normally work within?

5. An engineer designing an engine must take into account certain aims that might tend to work against one another, such as efficiency and performance, mileage and power, or economy and pollution levels. What skill or ability is he or she exhibiting when each of these aims is given its due?

6. What skill or ability is an engineer exhibiting when he or she succeeds in completing all engineering work within budget?

7. What skill or ability is an engineer exhibiting when he or she does a good job of gathering information needed for a project from several employees of a client and also contacts, and wins over to his or her solution, several of the client's decision makers?

8. List at least six skills not yet mentioned in this quiz that good engineers must develop.

1. Problems may be defined broadly and the assignment may be filled with ambiguities, but the engineer's solutions must be specific and explicit.

2. A process engineer is an engineer concerned with the methods and materials of manufacturing the items designed by engineers.

3. A project engineer is a member of an engineering team that coordinates the work of the team members.

4. The primary economic restraint within which a process engineer must work is the cost per unit of production and its impact on the profitability of a proposed product.

5. When an engineer succeeds in giving each project aim its due, he or she has probably exhibited the skill of balancing or handling conflicting objectives very well.

6. Working within budgets is to exercise the skill of operating within economic constraints.

7. Gathering information from a client's employees and convincing a client's decision makers involves the skill or ability to deal with people and their problems.

8. Other skills that a good engineer must develop over the years include the following:

Problem solving

Preparing, conducting, and evaluating experiments

Reaching good conclusions from a few observations

Finding, and eliminating, potential sources of error

Using information resources

Reasoning and analytical skills

Mathematical skills

Verbal and communication skills

Graphics skills, such as drawing, sketching, and producing graphs

Computation skills, including the ability to work with computers

Learning Segment 2
WHAT IT OFFERS

LEARNING OBJECTIVES

- Recognize some of the great variety of opportunities for employment in the field of engineering and its related activities.

- Attune yourself to some of the great variety of challenges open to those working in the field of engineering and its related activities.

OPPORTUNITIES

What, besides a respectable living, does an engineering career offer? Every individual answers this question differently, but they all agree that it offers certain rewards. The main one is the satisfaction derived from *creating*. Individuals have their own ways of expressing it: "something tangible results" . . . "a welcome outlet for my pent-up imagination" . . . "a thrill known also to the artist" . . . "visible solutions to real problems" . . . "the feeling is a good one." Each, in his own words, is telling about the joy that comes from creating useful structures, machines, and other complex physical contrivances.

Furthermore, this profession offers variety. There is a variety *within* each project; day-to-day activities include thinking, meeting, drawing, calculating, telephoning, corresponding, experimenting, searching in the library, consulting colleagues, and talking to customers. Similarly, there is variety from assignment to assignment. One project may involve designing a household water purifier, the next might be a fact-gathering assignment, the next might involve troubleshooting of a product already on the market, and the next the design of a water desalinization plant for a medium-sized city. Finally, there is variety in job opportunities.

Certainly one factor that makes engineering attractive is the uniquely broad range of career options. There are continually more sides to this profession; many types of work, many fields of specialization, and many problem areas. There are possibilities here to suit a broad range of talents and interests. One word sums it up: *variety*.

Many Specialties

There are the long-established, major branches of engineering, such as aeronautical, chemical, civil, electrical, industrial, mechanical, and metallurgical engineering. In addition, a number of new ones have evolved recently, such as those devoted to space travel and medical instrumentation. Furthermore, in practice, most engineers concentrate their efforts on one phase of a major specialty. For example, some mechanical engineers are experts in the design of mechanisms, others concentrate on refrigeration, and another group deals in propulsion systems. Thus, considering all the branches of engineering and the numerous subdivisions of each, it is apparent that there is a wide variety of specialities from which to choose.

Many Types of Activity

Within any specialty, there is a broad range of activities to choose from, defined by the following extremes. At one end of this spectrum is development engineering, the frontier type of job; the problems generally call for novel or technically advanced solutions, as in VTOL development work. This type of activity demands good training and continued professional updating in science, mathematics, and advanced engineering. Engineers who are happy and successful in this type of job thrive on a steady diet of complex technical problems.

In contrast to this is what is commonly referred to as sales engineering, involving a minimum of technical innovation but a maximum of involvement with people. Superior salesmanship, a keen interest in people, and a pleasing personality are highly desirable for this type of activity.

Between these extremes, there are hundreds of jobs that differ in day-to-day activities, challenges, and technical demands. They suit a broad range of abilities and preferences. All jobs over this broad spectrum involve problem solving; all involve working with things and with people; all require salesmanship; and all require creativeness and the other qualities described in Learning Segment 1, but in varying degrees.

CHECK IT OUT

Are there any areas of activity related to your job that you have not been directly involved in, but would be interested in pursuing? What can you do about it?

Many Types of Enterprises

Among the traditionally heavy employers of engineers are the aircraft, appliance, automobile, chemical, communications, construction, electronics, energy, metals, machinery, and transportation industries. The tempo of innovation in products, services, and manufacturing processes is accelerating, so that the demand for engineers in these industries continues to increase. This is true, by the way, even of the technical enterprises that you may feel have long since become static. Consider bridge design and construction, which could be cut and dried and therefore a relatively dull kind of engineering activity. It is quite the contrary. Innovation is conspicuous in this field (Figures 21 and 22); imaginative efforts to reduce cost, enhance appearance, and provide variety could provide the bases for a fascinating book. The point of this is that industries you might write off because you assume they are technically stagnant, in most instances do offer many challenging opportunities. Then there are the newer "glamour" industries (e.g., those associated with space and computers) that have become major employers of technical people.

Figure 21. This unique bridge across the Severn River in Great Britain is a significant departure from the familiar suspension bridge. The deck and what for a conventional suspension bridge would be its supporting truss are an integral unit, forming one large, continuous, hollow beam. The cross-sectional shape of this box beam, apparent in the photograph, is aerodynamically superior to the conventional rectangular truss. It has reduced the effects of cross winds by two-thirds and, as a consequence, significantly reduced the steel required throughout the structure. Each section of the deck, like that shown being hoisted into position, is fabricated on shore and then floated to the site. Adjacent sections are welded together to form one 990-meter beam, which becomes the main span. These and other innovations in design and construction methods resulted in a bridge costing about one-fourth as much as a conventional suspension bridge of comparable size.

FIGURE IT OUT

Assume that you have been asked to design a structure that will permit vehicular traffic and trains to traverse a body of water too wide for a suspension bridge and too deep for a trestle bridge or tunnel under the bottom. Can you think of any possible approach to this problem?

____

____

____

____

Don't feel bad if this one stumped you. The solution is quite imaginative.

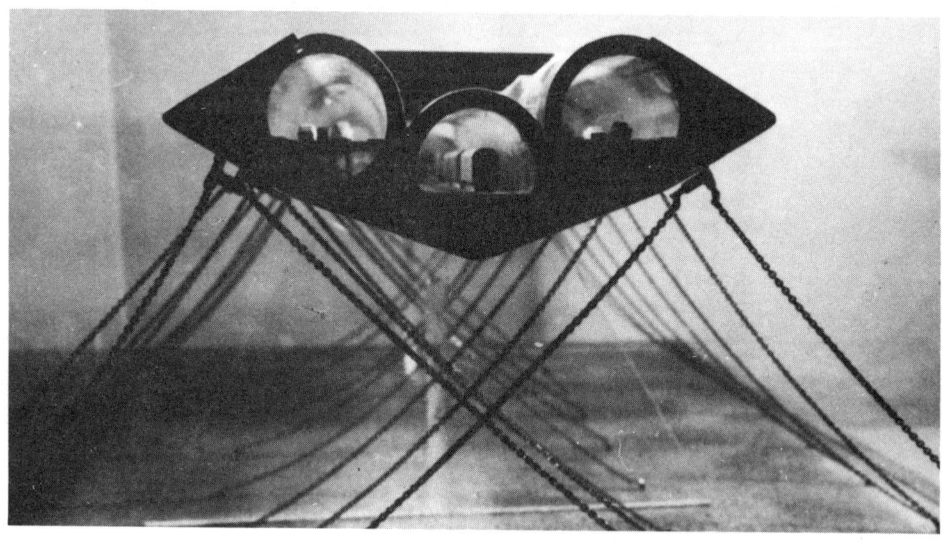

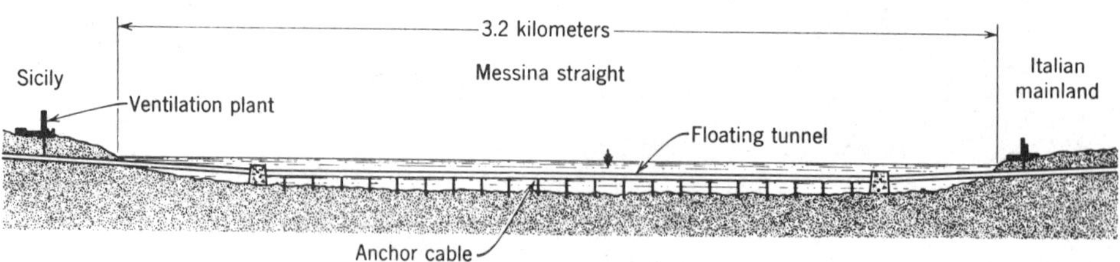

Figure 22. A bridge is not the only type of fixed crossing. In addition to the tube-in-trench concept employed in the Chesapeake Bay Bridge-Tunnel, there is the bored tunnel and the innovation modeled here—a floating tunnel (floating bridge, if you prefer). This new type of underwater crossing, proposed to connect Sicily to the Italian mainland, is especially attractive since the strait is about 3.2 kilometers wide—a little too wide for a suspension bridge. Furthermore, shipping traffic is such that a bridge is undesirable, the water is so deep that a bored tunnel is impractical, and frequent earth tremors in that area make bridge and bored tunnel unsafe. This floating tube presents no hazard to shipping, it is 46 meters below sea level, and it is relatively unaffected by seismic shocks.

CHALLENGE

The Challenges in the Area of Natural Resources

Many of the opportunities, much of the excitement, and a large share of the challenges awaiting engineers relate to the matter of resources. That covers a lot, including human, financial, information, material, food, water, and energy resources. For any one of these resources, the engineer may be involved in discovering sources, developing conversion methods, improving utilization, or developing means of recovering the expended resource.

Energy As a Case in Point. Surely you are generally aware of the rising demand, shrinking reserves, and other gloomy dimensions of man's energy problems. But you may not be so familiar with the variety of fronts on which engineers can contribute solutions and, therefore, where you might become involved.

THINK IT THROUGH

You should be able to anticipate a number of the areas of challenge in relation to natural resources. Try to list a few.

____

____

____

____

____

____

____

____

Answers to this exercise appear in the following paragraphs.

Sources. The basic sources of energy have been known for a long time, so nothing can be done about developing new sources. But the story is quite different in a closely related area. There will be a lot of activity in the search for new *supplies* for known sources of energy (e.g., new oil, coal, and uranium deposits). This includes the development of more effective means of locating underground and underwater deposits, as well as the search itself.

Furthermore, you may have observed that what will shortly be depleted for a number of resources are not the supplies but the *readily accessible* supplies. We are being forced to turn to difficult and expensive deposits of oil in the Arctic, oil mixed with shale, and oil beneath thousands of feet of ocean water. For instance, we have known about vast deposits of oil mixed with sand for centuries, but we never bothered with them—it wasn't necessary. Now we are interested. The development of technically, economically, and environmentally satisfactory systems for recovering oil under harsh conditions like those encountered at sea are better known and even more challenging (Figure 23).

Figure 23. The sea is one of the "new frontiers" for engineers, offering an impressive variety of challenges. Among them are: development of new types of ocean-going vessels, such as giant air-cushion freighters capable of 100 km/hr speeds; development of systems for the recovery of important minerals from the ocean bottom; and the design and construction of oil-drilling platforms like this rig shown headed for the turbulent North Sea. This type, known as a semisubmersible, has a platform area of 2800 square meters and houses 50 workers. It will be towed to the drill site, partially submerged, and then anchored until its work is done there; then it will be refloated and moved to the next site.

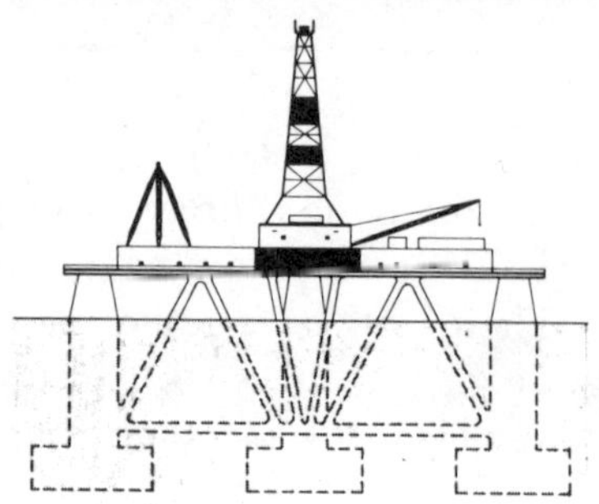

Conversion. There are a number of known but, thus far, virtually useless sources of energy. What's desperately needed are economical methods of converting the energy potential of these sources into useful forms. This is where a lot of energy research and development funds will be applied and, therefore, where much of the action and many of the jobs will be. Large-scale, economic methods of converting the energy potential of sunlight, wind, and refuse are in various stages of development; some are near success (Figure 24) and others are far from it.

Figure 24. You might have trouble recognizing it as such, but it's a windmill at the mouth of a large wind tunnel. This is a vertical axis type, with blades of balsa wood coated with fiberglass. It is not only simple and cheap, it is multidirectional (horizontal axis designs have a problem in this respect). This windmill, producing enough power to supply a typical household, is one of a variety of designs under development around the world.

Here is an excellent case in point. The theory of generating electricity by an ingenious method that capitalizes on the temperature differential between surface water and the depths of the ocean goes back almost a century (Figure 25). It's a beautifully simple concept, appealing indeed. But can it be developed into a practical conversion scheme?

THINK IT THROUGH

Before reading the caption accompanying Figure 25, examine the picture and the diagram and review the preceding paragraph. Then answer this question.

What basic energy source is this scheme attempting to convert into useful energy?

____

____

The answer to this exercise appears in the caption accompanying Figure 25.

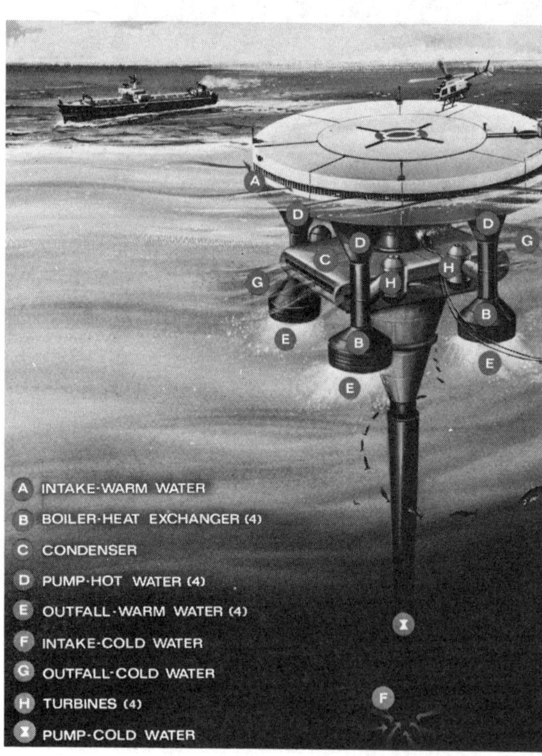

Figure 25. Much of the sun's energy reaching the earth is absorbed by the oceans. One scheme for tapping this energy closely parallels that employed by the steam turbine used in power plants. The main difference is that the "working fluid" is ammonia instead of water. Fortunately, ammonia vaporizes at the temperature of surface water and condenses at the temperature of deep water. The low-pressure ammonia gas propels a turbine, which drives an electrical generator. The accompanying diagram traces the complete cycle. The enormous platform pictured here houses the necessary boiler, turbines, condensers, and pumps required for this process. It would require very little human attention for ordinary operation. Such energy platforms are still in the conceptual stage; there are numerous problems to be resolved.

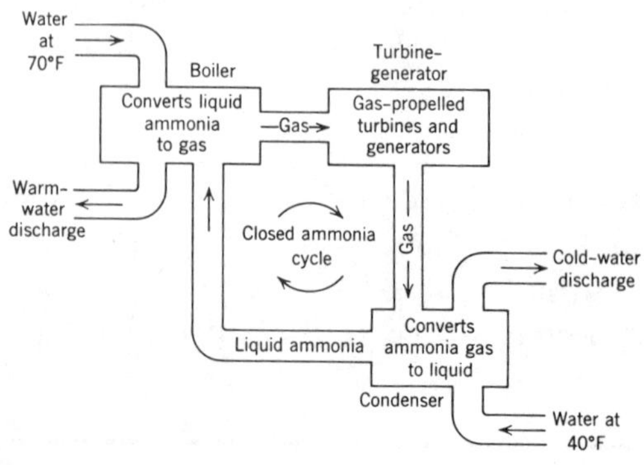

Storage. It has probably occurred to you how ill-equipped we are to capture and *store* energy from intermittent sources like sunlight, wind, and the tides. A breakthrough in energy storage capability would be of enormous technical, economic, and social significance. True, we have batteries, artificial lakes, molten salts, flywheels, and numerous other possibilities but, as of now, we are lacking a high-capacity, low-cost storage medium.

FIGURE IT OUT

There is one rather mundane, but very significant benefit that would result to our electric utility companies, and consequently to everyone, if a high-capacity, low cost storage capacity were developed. What is this benefit?

____

____

____

____

____

____

Such a capability offers more than the obvious payoff. We could obtain much better utilization of our power generating plants if electricity generated during the low-demand periods (the middle of the night, for example) could be stored and fed back into the system during peak-demand periods. The storage breakthrough we seek would produce the same effect as a 25 percent or more increase in power-generating capacity—nothing to sneeze at.

Utilization. Much of the work of an engineer involves resource allocation, and so the engineer is a major influencer of the rate at which man consumes a resource like energy. How could it be otherwise? The designer of a building determines the type and amount of energy the building and its occupants will consume over many decades. The designer of a steel mill does likewise, and so do designers of transportation facilities, communication systems, and agricultural equipment. Engineers determine not only what kind and how much, but the amounts of energy lost through friction, leakage, and the like, and how much "waste energy" is reclaimed and profitably used.

The growing pressure to conserve energy has added a new dimension to the work of all engineers whose creations consume energy. In many instances, like automobile design, it is a new era compared to the days when oil was considered abundant. To cite just one case, the designers of automobiles, trucks, and trains, from now on, will be expected to devote careful attention to the aerodynamic performance of their vehicles. Wind resistance significantly affects energy consumption, as Figure 26 illustrates. These added considerations in design mean new challenges and increased demand for technical talent.

Figure 26. This simple sheet metal "picture frame" attached to the leading face of a truck body illustrates how some attention to the aerodynamics of vehicle design can pay off. Trucks equipped with this inexpensive device have been found to use 23 percent less fuel than similar standard trucks at 90 kilometers per hour speeds. If all medium and heavy trucks were equipped with such devices, the fuel savings per year would be in the millions of barrels.

A Similar Story for Material Resources

You don't read about it as much as energy, but people, particularly in the metals, industries, have come to recognize material resources for what they are—limited. This has important consequences for engineers, because they have a lot to say about what materials will be used, in what quantities, with how much waste, with how long a useful life, and with what possibilities for recycling. Of course, it has always been their objective to minimize the material resources consumed by their creations without jeopardizing the users' safety. This is an important part of their job—to labor long and hard to "do the best with the least." A good example of how engineers can substitute ingenuity for materials is the clever scheme they devised to conserve the copper that goes into telephone wire. It is a device that enables a single wire to carry more than one telephone conversation simultaneously. Where the cost would be especially high, such as with transoceanic submarine cable, one strand carries more than a hundred conversations. You can imagine how many wires would be strung around the countryside if a separate strand were required between each two parties! Observe that engineers influenced the amount of a resource consumed—copper—through their design. Hopefully this will always be true, with even greater attention, care, and ingenuity in the future.

There are parallels between the problem areas associated with energy and material resources, with one notable exception. The opportunities for recapturing spent resources are excellent for materials, but negligible for energy. You are aware of the recycling process and its desirability, but you may not fully realize just how much the economic and technical feasibility of recycling are influenced by the *designer* of the product. This gives rise to a major challenge: to significantly increase the proportion of our natural resources that can be reused through careful forethought during design. To be sure, designing recyclability into products adds another dimension to design.

Other resources too. Don't let the relatively few words devoted to resources other than energy and material resources mislead you. Elaborating on all the high-opportunity areas is impossible. Energy and materials were chosen as representative examples of what is true for other resource problems. You read the newspapers, so you are aware of food and water shortages. You can take it for granted that a lot of imagination and plenty of technical manpower will be needed to alleviate such problems. Figure 27 indicates the kind of thing that has been accomplished.

THINK IT THROUGH

Perhaps you know what the large disk-shapes in Figure 27 (a) are or perhaps you have flown over them at high altitude and wondered what they were. Can you explain them without reading the caption first?

____

____

Figure 27. Without some help, you might be hard put to explain the disks pictured in (a). These are fields of alfalfa, barley, and wheat, each about one million square meters in area, made possible by irrigation machines like the one pictured in (b). Each machine pumps water from a well at the center of the circular field and distributes it through a pipe that sweeps as a radius every 24 hours. The 560-meter pipe is supported by a superstructure mounted on rubber-tired carriages spaced about 60 meters apart and driven by electric motors. (The speeds of these motors must be precisely synchronized so that adjacent sections remain aligned as the whole thing rotates. That is a challenge.) A series of such machines is gradually converting formerly useless African desert into croplands—presenting an odd but most welcome sight.

As an illustration of what might be done, consider the possibility that man may eventually embark on multibillion-dollar construction projects for redistributing fresh water over whole continents. (This has been proposed for North America and calls for distribution of Arctic fresh water through Canada, the United States, and Mexico in order to meet their requirements.) Although they may strike you as desperation measures, such schemes may come to pass; the problems are that serious.

CASE STUDY: A DOMESTIC WATER DESALTER

Because of the diminishing supply of fresh water and the rapidly increasing demand for it, the problem of providing adequate amounts of drinking water is a pressing one. Developing economical sources of drinkable water is an *engineering* problem.

In many areas of the world, a promising source of drinking water is brackish water that lies underground. In anticipation of both the commercial and the service-to-mankind opportunities in this activity, a company is developing equipment that will convert such water to a usable form. One of these projects is the development of a converter that can be used in the home. The engineer responsible for this is now evaluating a promising prototype (Figure

28); it both purifies and desalts the water. The input is salt, brackish, or otherwise impure water; the output is demineralized and potable. Such a converter would be useful in homes; small commercial establishments, military field units, and on ships. It is effective and simple, needs little maintenance, and does not require a pressure vessel.

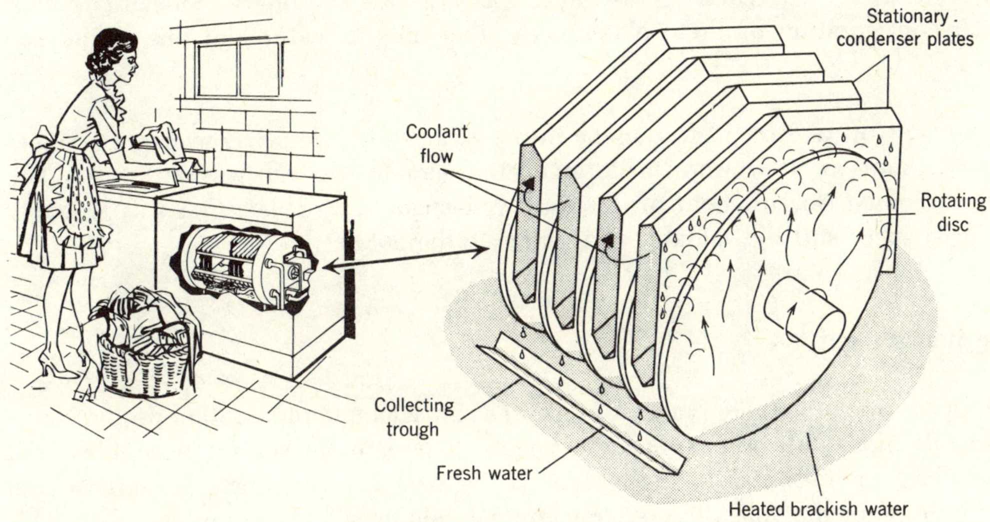

Figure 28. Household unit for converting brackish water or seawater. Close-up view shows the actual conversion mechanism. As the shaft rotates, the discs connected to it are coated with a thin film of warm, brackish water at the bottom of the tank. This thin film of water vaporizes as it passes through the air and condenses on the cool stationary plates. This condensate, now fresh water, drops into the collecting troughs.

THINK IT THROUGH

Basically, all that is needed for converting brackish water to fresh water is a distillation process. So what's the big problem all about?

____

____

____

____

An answer to this exercise appears in the following paragraphs.

The problem is not simply a matter of finding *a* method of converting brackish water to potable form; after all, distillation has been known for centuries. The problem is to find a means of transforming *large quantities* of water at a *reasonable cost*.

The development of this converter is based partly on the engineer's technical and scientific knowledge and partly on inventiveness. He could not have developed this machine without understanding vaporization and condensation phenomena, the behavior of thin films, thermal processes, and other scientific facts. This knowledge alone, however, could not have yielded this converter. The rotating discs interspersed with stationary collector plates, the particular configuration of these plates, and other unique features of this mechanism are *inventions*.

Real engineering effort and talent have gone into this machine. Many conversion schemes were evaluated, hours of tests were made, and considerable research was necessary. The result of this extensive development process is a well-engineered device that will probably prove financially successful and valuable as a service to the public.

Solid-Waste Management

This is one aspect of the broader matter of resource management and another problem area that will provide many future jobs, in two respects. One is in *developing* new collection, sorting, recycling, and disposal methods and means of generating energy from material waste. The other is in *applying* these systems on a community-by-community basis. This is where most of the jobs in the waste management field will be found.

There is a parallel between environment-related opportunity and solid-waste management in that (1) you are generally aware of the nature and magnitude of the problems, (2) there is need for imaginative technical talent to develop new pollution abatement and control capabilities, and (3) the majority of jobs will be not in development but in application of abatement and control systems to steel mills, power plants, refineries, and other sources of pollution.

Potential Contributions in the Health-Care Field

A sampling of recent achievements indicates what engineers might contribute in the future. The modern operating room is filled with impressive-looking machines that are routinely employed. There also are accomplishments such as the heart-lung machine, the miniature devices for incisionless internal surgery, and the equipment for neurosurgery by freezing.

For medical diagnosis there are tiny instruments that can be inserted into vital organs (including the heart through a vein) to make measurements, apparatus that automatically analyzes blood samples, and computerized analysis of electrocardiograms. Other diagnostic innovations are in the developmental stage.

Prosthetic devices are numerous, indeed. There is the miniature battery-operated heart pacer that is embedded in the patient's body to keep his heart operating by sending electrical pulses directly into the heart muscles. A number of artificial organs and vastly improved artificial limbs are under development. The most dramatic are the artificial kidney and the mechanical heart.

Exciting developments are taking place in hospitals. A variety of ingenious sensors are being developed for continuous measurement of patients' temperature, blood pressure, and so forth, so that these vital signs can be monitored and an alarm sounded when a patient's condition warrants attention.

CHECK IT OUT

You may have had some experience with health care delivery systems. Do you have any ideas about areas of challenge for engineers in the field of health care?

____

____

____

____

____

Your answer to this exercise depends on your experience and only you can judge the appropriateness. However, you might enjoy comparing your list with the several suggestions that follow.

In spite of recent developments, some major challenges remain. Here are some of them.

- Since hospitals are so highly instrumented, they need "resident engineers" to oversee the acquisition, installation, and care of this complicated equipment, and particularly, to protect the safety of persons coming in contact with it, in view of the high voltages or radioactive materials that are often employed. All hospitals need such technical advisors, but few have them.

- There is a need for the continued development of diagnostic aids (specialized instruments and computer programs) that will make diagnostic services available at low cost to masses of people. Machines that quickly and cheaply analyze test specimens, electrocardiograms, X-rays, and other diagnostic evidence will significantly advance the prevention and early detection efforts of physicians.

- The information-handling problems of a hospital are approaching the impossible. Information handling includes gathering, storing, distributing, processing, and retrieving information on patients. You wouldn't believe the cost of all this. Nor would you like to hear about all the misdirected, misplaced, transposed, and too-late information, nor the gruesome consequences of these slipups. The computer will be a boon in these respects, but people are needed to do the applying which, on the basis of attempts so far, appears to be especially challenging.

• There is the perennial problem of making inventions widely available at reasonable cost. For every inventor of an artifical kidney, hundreds of engineers are needed who are willing to refine the original and usually very costly machine in order to reduce the cost to a tolerable level. This requires persistent, imaginative technical effort, which is absolutely essential if a significant number of people are to benefit from the basic idea.

This sampling gives you an idea of the engineering opportunities available in the field of health care. Incidentally, there are some established specialties within the field of "medical engineering." The *biomedical engineer*, for instance, focuses on the development of instrumentation for medical research, diagnosis, treatment, care, prosthesis, and education (Figure 29).

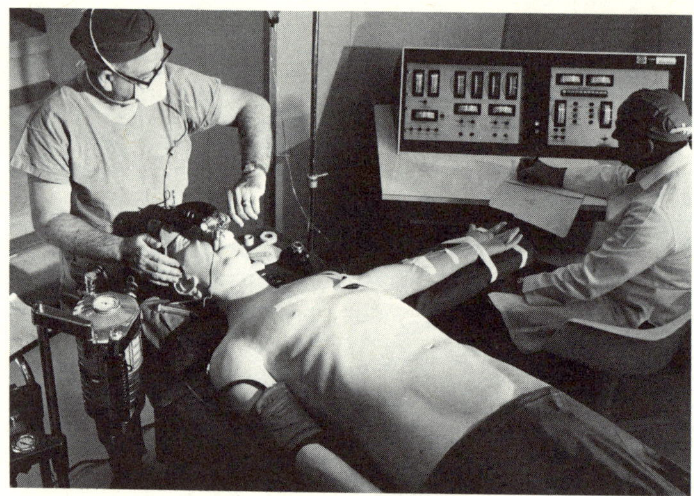

Figure 29. This is a computer-controlled patient simulator for training anesthesiologists, developed by engineers. It "breathes, coughs, and vomits," has a "heartbeat," and can "die." In these and many other respects Sim One closely mimics human behavior. The student can give injections, administer oxygen, check pulse rate and pupil dilation, and perform most other routine and emergency functions of an anesthesiologist. The instructor, using the console at the right, can program the lesson, evaluate performance, temporarily stop the process to discuss a point with the student, and have a lesson repeated if performance is not satisfactory. Obviously, these are privileges not available when real subjects are used. This simulator significantly shortens the training period for an anesthesiologist and allows instructors to exert much closer control over the learning process.

There are a number of programs in biomedical engineering. The *clinical engineer* is also concerned with instrumentation, prosthesis, and the like, but on a patient-by-patient basis. He works as a member of a medical team, supplying the expertise needed when, say, a special device must be developed to be implanted in a particular patient. The *hospital engineer* is a technical advisor and overseer, whose responsibilities were outlined earlier. Notice that although there are some emerging subspecialties in the field of medical engineering, opportunities in the field are certainly not restricted to those labeled subspecialties.

Summary

Because natural resource and health care problem areas have been emphasized, do not conclude that "that's it." These illustrate in some depth what is to be found in many problem areas.

An analysis of the problem areas in which the engineering profession stands to make significant contributions reveals a special need for engineers who offer a *broader perspective*, a *greater imagination*, and a *higher sense of purpose* than have prevailed up to now. On the matter of broader perspective, interrelated problems are too often attacked separately, as is the case with the many interdependent problems that plague a city. Often a solution to one problem aggravates other problems. Engineers are needed who will view a situation with broader perspective than has been customary and produce a well-integrated solution to the whole problem.

There is need also for engineers who will bring fresh ideas to the many long-standing, still unsatisfactorily solved problems. Society can use engineers who are bent on *imaginative* application of their knowledge and skills.

Engineering also needs many more practitioners dedicated to applying their talents where society's needs are greatest. The engineering profession and its related occupations can contribute toward easing man's resource plight, improving the environment, extending low-cost medical care, combating starvation, alleviating conditions that breed social ferment, aiding the physically disabled, improving travel safety and mass transportation, converting wastelands into croplands, and aiding underprivileged nations.

SELF QUIZ

1. The fact that there are many specialties within the field of engineering contributes to the great variety of opportunities. What are two other characteristics about employment in the engineering field that contribute to this variety of opportunities?

2. There is an unlimited number of challenges for engineering and related activities. List some challenging problems to be solved in the area of energy resources.

3. Define the term *biomedical engineer*.

4. Define the term *clinical engineer*.

5. Define the term *hospital engineer*.

1. Engineering provides opportunities as an occupation not only because there are many specialties, but also because there are many different kinds of activity from day to day within each given field and because there are many different kinds of employment situations, from large corporations, government and private sector activities, small to medium sized manufacturing and consulting firms, many small to medium consulting firms, and self-employment as a consulting engineer.

2. Some challenging problems in the area of energy resources are

Seeking, locating and identifying, and utilizing new sources and supplies of energy

Efficient methods of conversion of such basic energy sources as sunlight, wind, tides, nuclear, and others into useful and beneficial energy

Discovering a high-capacity and low-cost storage capability for useful energy

Designing efficient utilization of energy into nearly any product, tool, process, or installation (including the effective trapping of "waste" energy)

3. A biomedical engineer focuses on, or specializes in, the development of instrumentation for research, diagnosis, care and education in health care fields.

4. The clinical engineer is concerned with the same things that a biomedical engineer is, but in this case, the specialty focuses on the individual patient's needs.

5. A hospital engineer is a technical advisor and overseer that helps health care institutions such as hosptials acquire, install, maintain, and safety utilize the complex equipment now used in advanced health care facilities.

ISBN 0-471-01702-7